Building an Information Technology Industry in China

National Strategy, Global Markets

James A. Lewis

Deng Xiaoping's decision to open China's economy to foreign investment changed the world's information technology industries. China had a negligible presence in world markets for information technology (IT) in 1990. Its information technology sector has now moved to center stage, exciting admiration, envy, and fear. China is a leading center for the assembly of IT products, and its political and business leaders are eager to move on to more valuable activities. The last five-year plan announced the ambitious (and probably unattainable) goal of making China's IT industry "comparable to those in the United States and Japan."[1] China is pursuing a lead position in semiconductor manufacturing and is attempting to build a globally competitive software industry. More important, China is becoming a prime location for research and development (R&D) in information technology, often in partnership with foreign firms.

This report looks at China's efforts to build a national IT industry. It examines in particular the work in semiconductors and software, the two engines of IT, and asks what led to accelerated growth and whether it is sustainable. It asks how much of this success can be attributed to government policy and how much to external factors. The combination of government support, privatization, and the removal of barriers to foreign participation has been crucial for China's IT success. But the degree and nature of this success can be easily misunderstood

---

[1] "Summary of the Tenth Five-Year Plan (2001–2005)—Information Industry," http://www.trp.hku.hk/infofile/china/2002/10-5-yr-plan.pdf.

1

when China's performance is taken out of its global context. Data from the OECD showed that China briefly overtook the United States as a high-tech exporter in 2005, but the Chinese government reported that 90 percent of these exports were made by foreign firms operating in China. China's IT sector is large and growing, and several Chinese companies are gaining a significant share of international markets, but this success is not the conquest of a global industry but rather being absorbed by it.

China's leaders in the 1980s began to lay the groundwork for a modern IT industry. They did this for reasons of prestige, economic growth, and national security. They believed that China would be more secure if it did not depend on foreign IT products and could instead use IT products made entirely in China. This report shows why their goal will be difficult to attain in an increasingly integrated and globalized industry.

A National IT Industry in China

China used a combination of preferential policies to accelerate growth in the IT sector. Five-year plans and other guidance from Beijing provide a framework for local and provincial governments to decide how to allocate resources and compete to attract IT investment. The factors that explain growth are:

- A decision by China's leaders in the 1980s to pursue policies that opened the Chinese economy to foreign participation, reduced direct state control of enterprises, and expanded opportunities for entrepreneurship;

- Heavy direct and indirect subsidies, including direct financial subsidies, tax and trade concessions, grants of land, investment in infrastructure, and government support for graduate education and for research and development;

- A broad national effort to extract concessions and technology from Western companies in exchange for market access;

- Trade policies that provided, at least until accession to the World Trade Organization (WTO) in 2001, powerful incentives for companies to move to China and supported technology transfer efforts;

- The attraction of access to a giant and fast-growing new market for a broad range of IT products;

- Access to the skilled science and technology workforce that government investments in science and engineering programs have produced.

Each of these factors played a critical role in reshaping China's IT industry, but the most important were the decisions to relax government control and open China's economy to foreign participation. If China had decided to continue an industrial policy where government interference in firms' decisions remained high, its IT sector would still be uncompetitive and backward. The decision to

Building an Information Technology Industry in China

National Strategy, Global Markets

A Report of the
Technology and Public Policy Program
Center for Strategic and International Studies

Author
James A. Lewis

May 2007

About CSIS

The Center for Strategic and International Studies (CSIS) provides strategic insights and practical policy solutions to decisionmakers committed to advancing global security and prosperity. Founded in 1962 by David M. Abshire and Admiral Arleigh Burke, CSIS is a bipartisan, nonprofit organization headquartered in Washington, D.C., with more than 220 employees. Former U.S. senator Sam Nunn became chairman of the CSIS Board of Trustees in 1999, and John J. Hamre has led CSIS as its president and chief executive officer since 2000.

CSIS experts, working through more than 25 programs, conduct research and analysis and develop policy initiatives grouped under three themes:

Defense and Security Policy. Currently devoting over a third of its resources to security issues, CSIS has chosen to focus on both the traditional drivers of national security—defense policy and organization—and some of the most important new dimensions of international security, such as post-conflict reconstruction, proliferation, and homeland security.

Global Trends. A growing and more mobile population and advances in economics and technology have exposed the inability of nationally organized governments to respond to transnational challenges. A large number of international organizations, multinational corporations, and nongovernmental entities now exert significant influence over international affairs. CSIS examines not only how traditional nation-states themselves deal with problems that cross national boundaries, but also how they relate to this new and powerful group of actors.

Regions. CSIS is the only institution of its kind with resident programs on all the world's major regions. The programs enable CSIS to anticipate developments in key countries and regions across the world, especially as they affect global security.

CSIS does not take specific policy positions; accordingly, all views expressed herein should be understood to be solely those of the author(s).

Library of Congress Cataloging-in-Publication Data

Lewis, James Andrew, 1953–
  Building an information technology industry in China : national strategy, global markets / James A. Lewis.
     p. cm.
  ISBN-13: 978-0-89206-489-2 (pbk. : alk. paper) 1. Computer industry—China.
2. Computer industry—Government policy—China. I. Title.
  HD9696.2.C462L49 2007
  338.4'70040951—dc22   2006025622

The CSIS Press
Center for Strategic and International Studies
1800 K Street, N.W.
Washington, D.C. 20006
Tel: (202) 887-0200
Fax: (202) 775-3119
Web: www.csis.org

Contents

open the economy to foreign participation provided financial and intellectual property resources. Without foreign participation, China's IT industry would have remained undeveloped. The creation of "private" companies (rather than state-owned enterprises) was also crucial. Although many private firms in China have strong ties to government ministries, the shift in management to "private" and away from "government" created more efficient producers and opened greater opportunities for partnership with foreign entities. Finally, as economic openness and "privatization" produced greater wealth across the economy, domestic demand for IT products expanded, helping to make China an attractive destination for foreign IT investment.

The attraction of the China market, which offers hundreds of millions of new consumers, is now cited by Western companies as one of the primary reasons for foreign investment in China's IT industry, replacing the old view of China solely as an IT export platform. They report that the rate of growth in demand for IT products in China is much faster than in mature markets. China's market accounts for 10 percent to 15 percent of sales for many of the larger multinational IT firms. The impression among foreign IT companies is that investment in China will still yield greater returns because China's domestic demand for IT products is the fastest-growing in the world.

The opening of China's economy to foreign participation must be viewed from the perspective of the changes in the international economy that began with the end of the Cold War. The acceleration of economic internationalization and market integration that began in the early 1990s reinforced the effect of China's policies. These changes helped create a powerful surge in foreign investment around the world at a time when China was making itself a more attractive destination for investors. China's economic opening also came when many companies sought to concentrate on "core competencies" and to outsource functions where the return on investment was low. Shedding low-end, low-value activities (such as commodity manufacturing) or relying on contractual arrangements for production became a successful strategy for business. Foreign firms were quick to take advantage of China in implementing these strategies and incorporate China into global supply chains.

Certain global IT industry trends, like disaggregation, have also accelerated growth in China's IT sector. The industry, which has been moving away from vertical integration for some time, uses disaggregation as a strategy to increase competitiveness. Companies focus on activities where they are more competitive or where the returns are greater. They also reduce their costs by outsourcing activities where fixed costs are high. Reductions in the cost of transportation and communications allow companies to spread design and manufacturing tasks around the world—to their own subsidiaries or to trusted suppliers—and take advantage of cost differentials in different countries. The result is a manufacturing process where it is normal to design and manufacture IT products and components in several countries and assemble them in yet another country, none of which may be the home country of the company whose label goes on the shipping container. American IT companies see this global supply chain as a source of strength, and their incorporation of China's production resources into their supply chains has been a means to gain competitive advantage.

An active presence in China is now a key part of major IT corporations' strategy for global competitiveness. Companies believe that a presence in China will improve their ability to develop products better suited to the China market. Partnership with a Chinese firm or institution can also be an important tool for successfully navigating the many formal and informal requirements imposed by Beijing and local governments.[2]

China's national, provincial, and local governments woo foreign IT companies by offering a broad range of incentives that would be the envy of any American state or European nation that has tried to attract a plant or company. Regional governments compete with each other to attract foreign investment. Large American IT firms describe China's efforts as similar to the techniques used by states in the United States or provinces in Europe—essentially a competition among governments to offer the most attractive package of incentives.

Incentives include low-cost or no-cost financing, land grants in the form of long-term leases, construction loans, tax and trade credits or rebates, the creation of special export zones, and investments in electrical power and other utilities, communications, and transportation infrastructures. According to interviews with companies that have invested in China, some of the incentives can be quite large—one program offers a 40 percent tax rebate if profits are reinvested in China for five years.[3] The city of Ningbo alone set aside $1.2 million in 2004 to support local software companies. A Ningbo official said, "We will offer more favorable policies and a good investment environment alongside strengthening the existing information industry" in order to attract foreign investment and high-tech companies. Ningbo is already home to Ningbo Bird, a fast-growing and successful competitor in the cell phone market.

Tax and trade concessions reinforced subsidies. Chinese trade policy, especially before WTO accession, gave an advantage to firms that had a presence in China. Companies that located and manufactured in China faced lower tariff hurdles than those located outside. Most companies cite relief from import barriers as a major factor in their decision to locate in China. Regional governments used offers of trade and tax exemptions to compete with other regions of China and with other countries to win foreign investment.

A Taiwanese investor in a major semiconductor facility described the accommodations that the Shanghai government was willing to make to win his company for their city: Shanghai offered land at a no-cost, long-term lease. It made investments in infrastructure (such as roads) to improve the property. It arranged for the provision of low-interest lines of credit and waived taxes and fees

---

[2] Ramesh Adhikari and Yongzhen Yang, "What Will WTO Membership Mean for China and Its Trading Partners?" International Monetary Fund, *Finance & Development* 39, no. 3 (September 2002), http://www.imf.org/external/pubs/ft/fandd/2002/09/adhikari.htm.

[3] "Sights Set on High-Tech Industry," *China Daily,* May 13, 2004, http://service.china.org.cn/link/wcm/Show_Text?info_id=95258&p_qry=informatization.

for the first few years. It provided training subsidies for the company's employees and expedited visas for foreign employees. Permission was granted for the company to have its own schools for employees' children and even open a church on the facility grounds. Although the investor still had to arrange the bulk of the financing for the manufacturing facility in Western equity markets, the subsidies provided by the Shanghai government greatly reduced the risk of failure in the critical early years.

Accession to the WTO means China can no longer use the trade incentives it once offered to attract foreign investment. They are banned by WTO rules. The effect of this will be to change the nature of foreign investment in China (fewer joint ventures), but the loss of trade incentives will not cause foreign firms to stop investing in China. The timing of China's accession was such that a desire for access to China's human capital and domestic market and the lowering of risk for doing business in China as it adopts international norm will compensate for WTO-related change.

Although WTO accession has reduced the benefit of some incentives, companies now cite a new reason for going to China—the growing attraction of China's large, educated workforce. This factor is continuing to drive foreign IT investment. The science and technology workforce that attracts foreign companies is the result of government investment decisions made two decades ago.

In the mid-1980s, in reaction to new U.S. and European R&D programs, prominent Chinese scientists wrote to Deng Xiaoping, saying that China was falling behind industrialized countries in developing the advanced technologies that would be the key to national power. This letter led China to create the High-Tech Research and Development (863) Program, which invests in high-tech R&D projects to improve Chinese security and international competitiveness. The effect of the increased investment in R&D has been to increase the number of graduates with a science and engineering background.

Government spending on education and R&D through programs like 863 or the Ministry of Science and Technology's Torch Program provides an incentive for foreign firms to locate in China. Government investments in education and training (i.e., human capital) are an indirect subsidy that makes China an attractive destination. China now has a relatively large pool of engineers and programmers available at low cost. NASSCOM, the Indian software manufacturers association, has estimated that China has more than 100,000 software engineers earning salaries that are, on average, half of what their Indian counterparts earn. The research projects that 863 funded are less important than the human capital the project created.

China's motives in creating the 863 Program were to increase national prestige, increase China's economic competitiveness with other nations, and improve national security by building a strong economy, developing the means to build high-tech weapons, and reducing reliance on suspect foreign products.[4] The 863 Program funds R&D efforts at Chinese research institutions and universities (and now at Chinese high-tech companies) in 19 technology areas, including information technology. Its principle for selecting projects is to "combine military and civil use, with emphasis on the latter."

Most research institutions funded by 863 were affiliated with the Chinese Academy of Sciences (CAS) or with the People's Liberation Army (PLA). CAS alone has more than 100 affiliated research institutions scattered throughout China. This investment, a legacy of Soviet-style economic planning, resulted in a large, hierarchical structure of research institutions around the country. Since the 1990s, both CAS and the PLA have pushed to privatize and commercialize their research, a move that has greatly increased their contribution to the IT sector and that appears to mimic the U.S. experience with universities like MIT and Stanford. The large investment in research and education that began in the 1950s means that China has an advantage over many of its potential IT competitors in the developing world as a result.[5]

Initially, the IT-related R&D funded by 863 produced technologies that were neither advanced nor commercially competitive. This has changed in the last decade. The reasons for this are not only a maturing of Chinese R&D, but more important, a shift in how R&D is funded and managed. This shift reflects economic policy decisions to privatize, allow expanded foreign participation, and introduce market disciplines.[6]

Advances in China's IT industry came faster when government officials reduced their role in directing the industry, particularly after a State Council decision to "separate the government functions from those of the enterprise."[7] Research institutions were allowed to form privately managed companies, albeit with government agencies often as the principal investor. Leading IT companies like Legend (now renamed Lenovo), Great Wall, and Langchao were all spin-offs from government institutions.

These "privatized" organizations had key advantages when compared with their official predecessors. They were faster and better at commercializing their R&D, and they were better able to attract foreign partners and financing. Many 863-funded projects at government institutions and universities were "spun off," with government agencies and government-run banks playing the role of venture capitalists. The individuals from the companies and agencies involved in the new ventures have a complex mixture of private and public motives. They are

---

[4] People's Republic of China, United Office of Hi-Tech Program, Ministry of Science and Technology, "High-Tech Research and Development Program of China," http://www.863.org.cn/english/.

[5] Chinese Academy of Sciences, "Progress in the Initial Stage of the KIP Pilot Project," http://english.cas.ac.cn/eng2003/news/detailnewsb.asp?InfoNo=20964.

[6] United Office of Hi-Tech Program, "High Tech Research and Development Program."

[7] People's Republic of China, State-owned Assets Supervision and Administration Commission of the State Council, "China State-owned Assets Management System Reform Entering New Stage," May 22, 2003, http://www.sasac.gov.cn/eng/eng_qygg/eng_qygg_0001.htm.

pursuing a national program aimed at increasing China's national power and prestige, but they are at the same time competing for private gain, often against other government-supported efforts.[8]

Technology Transfer

"Future conflicts may well be competition for the possession of knowledge. Now all the most valuable intellectual property is in the hands of the Americans. That's not right."

Li Yongjun, Shanghai Science and Technology Bureau[9]

Technology transfer plays a central part in explaining the rapid growth of China's IT industry, but at the same time, we should not overstate its value. China is still dependent on foreign technology despite a decade of heavy investment and industrialization. Western firms report that technology transfer concessions are a part of every negotiation with the Chinese, and the blurred line between private companies and government ministries provides an advantage to Chinese firms. These transfers include technological "know-how" and business and managerial skills. Western executives say that one of the chief benefits to China of foreign investment is this transfer of crucial business and managerial skills.

China attempts to regulate technology transfers so as to maximize potential benefits. Its regulations can restrict the ability of a foreign company to make Chinese partners sign confidentiality agreements that would force them to guard technology or to restrict sales of derivative products made from proprietary technology. Western firms complain that these regulations, often buried in China's complex, overlapping, and opaque regulatory structure, can skew licit transactions in favor of Chinese firms.

China agreed to bring its rules into line with WTO protections for intellectual property (IP), but weak enforcement, powerful economic motives, and cultural predilection for imitation have meant continued problems for IP protection. Chinese companies or employees do not face serious risk of penalties from legal or government action for IP violations. Some Taiwanese and American business companies say that China will never be serious about protecting IP until it has IP of its own to protect. Illicit acquisition of proprietary technology by employees or partners remains a great concern for Western companies doing business in China.

---

[8] Report of the People's Republic of China, APEC TEL 18 WG Meeting, September 1998, http://www.apectelwg.org/apecdata/telwg/18tel/plenary/plen-m-13.pdf; for example: "Since 2000, the CAS has accelerated its work of transforming its institutions of technology development into business enterprises and establishing modern enterprise systems at its existing firms. By the end of 2001, the first batch of 13 CAS institutions had been turned into companies." Chinese Academy of Sciences, "High-Tech Industry Development," http://english.cas.ac.cn/eng2003/page/T46.asp.

[9] Allen Cheng, "Bargaining Chips," *Asiawee* 27, no. 5 (February 9, 2001).

Chinese firms now also complain about the theft of IP by employees, and stories of an ex-employee leaving to set up a competing business based on IP from their ex-employer are not unusual. Foreign pressure and WTO commitments have helped move China in the direction of stronger IP protection, and statements from Western and Chinese companies suggest that there is also a growing realization in Beijing and elsewhere that weak IP protection is a disincentive to innovation by the Chinese themselves.

A few Chinese firms have a clear strategy to illegally acquire their competitors' technology (although the same charge could be made against European and Asian firms). A HuaWei employee was caught at a trade show in Chicago surreptitiously dismantling competitors' equipment and photographing it. The HuaWei employee, who had made some effort to disguise his affiliation, was carrying a list of U.S. and Japanese firms whose equipment he wished to examine. HuaWei had previously faced charges from Cisco that it had "misappropriated" Cisco's IOS software and infringed Cisco patents. HuaWei was obliged to compensate Cisco for its actions. One executive at a Japanese technology company noted: "This sort of stuff, especially the stuff with Cisco's IOS, will eliminate HuaWei from doing business with all the major players in telecom like the RBOC's and [U.S.] Government."[10]

Western companies are aware of the risks to their IP from both government pressure to share and from unauthorized access by individual employees. Companies employ a number of stratagems to reduce the risk of IP loss. These include holding back key processes from Chinese employees, retaining advanced functions outside of China, installing monitoring or surveillance equipment, or allowing access only to low-end technologies. Fear over the loss of intellectual property is a clear disincentive for some investments and partnerships. An informal poll of 30 Western semiconductor firms found that 18 had decided against locating in China because of concerns over theft of IP. The remaining 12 firms that decided to locate in China had taken explicit steps to safeguard their IP, particularly design and manufacturing process information. Many IT companies cite risk to IP, along with regulatory uncertainty, as the two major obstacles to doing business in China. The result is that foreign investment in Chinese-owned firms has focused on production and sales, not research, as opposed to ventures where foreign control was greater (and IP risks correspondingly lower).[11]

China also has legitimate means of acquiring technology and know-how. These include research partnerships with Western companies and universities, sending Chinese students to foreign educational institutions, and attempting to attract Chinese expatriates back to the mainland to establish new firms. Over the long term, these avenues offer China more reliable access to technology, but in the

[10] Personal communication to the author, August 2004.

[11] Peter Marsh, "Fear of High-tech Piracy Makes Some Microchip Companies Cool about China," *Financial Times,* July 15, 2004; "Foreign Firms Contribute Little to Technology Spread: Survey," *People's Daily,* November 19, 2003, http://english.people.com.cn/200311/19/eng20031119 _128555.shtml; CNETAsia, "Foreign Investment Flunks R&D in China," November 21 2003, http://asia.cnet.com/news/industry/0,39037106,39158792,00.htm.

near term China will continue to pursue national policies to extract technology from foreign investors.

Companies say that they give their Chinese partners older technologies to which they assign a low value and keep high-end technology to themselves. One of the largest Japanese IT manufacturers even found a profitable sideline in selling to China the equipment it is retiring as obsolescent. This firm and others gamble that the pace of technological change is such that the technology Chinese firms acquire today will soon be outmoded. In many cases, this assessment has proved to be accurate. However, since transferring low-end technology is an iterative process that is repeated every few years, the level of technology provided to China continues to rise. In this sense, China is following a pattern seen earlier in Japan, Taiwan, and Korea (and in the United States during the nineteenth century). Taiwanese executives see a reflection of their own experience in building an IT industry, as Chinese companies use these older technologies as a bootstrap to pull themselves into higher-value manufacturing.[12]

China's leaders are very keen to move away from this foreign dependency. They want to see China become a leading global source of innovation. The importance of innovation for growth has been a recent theme in economic policy literature, and this has been reflected in policy statements in the United States, the European Union, and many other countries. Innovation is particularly important for China, as its economy is relatively inefficient. For all its manufacturing strength, China's workforce is remarkably unproductive compared with the workforces of the United States, Europe, Japan, or Korea. Government figures estimate that although GDP has grown by 9 percent per year for the past few years, only about 1 percent has been due to productivity growth. The rest has come from adding more inputs, not from using those inputs more effectively. China is inefficient in its use of resources—it takes far more energy to produce or transport something in China than it does in other developed nations. China hopes to innovate its way out of these problems. The problem is that the pace of innovation can be slow when intellectual property protections are weak.

There is also a concern that China is only manufacturing other people's products, not coming up with products and ideas of its own. This worries Chinese leaders more than they will admit in public. They do not like being dependent on foreign intellectual property. Some Chinese leaders in business and government see China's dependence on foreign IP and standards as another form of colonialism that they must overcome. When AMD, an American semiconductor manufacturing company, decided to transfer an older generation of CPU technology to China, a few Chinese even complained that the deal only reinforced China's dependence on foreign ideas.

---

[12] Douglas Fuller, Akintunde Akinwande, and Charles G. Sodini found that firms unloaded "mature technologies" that had declining profit margins. See "Leading, Following or Cooked Goose: Explaining Innovation Successes and Failures in Taiwan's Electronics Industry," MIT IPC Globalization Working Paper 01-001, http://ipc-lis.mit.edu/globalization/globalization%2001 -001.pdf.

The emphasis on innovation poses something of a challenge for China. A state-directed economy has advantages in some areas. It can command that resources flow to a specific area with the speed and commitment that a democracy can sometimes find hard to match. China has spent enormous resources creating the human capital for technological innovation in the last two decades. It has built a scientific research establishment that, in a few areas, is world class. Between 1995 and 2005, China more than doubled the percentage of its GDP invested in R&D—from 0.6 to 1.3 percent.[13] This is still small, but China says that it intends to double the proportion of its science spending devoted to basic research to about 20 percent of its science budget in the next 10 years. The Chinese share of the global production of graduate degrees in science and engineering, publications, and patents grows steadily every year.

But this investment in education and R&D is not enough for innovation. Institutional and cultural impediments slow innovation in China. One sign that Beijing recognizes this is the recent announcement that China's education system would adopt "American-style" textbooks and methods to teach science. These methods emphasize creativity and discovery rather than rote learning and memorization. The question for Beijing is how to create spontaneity and independence in scientific research while curtailing them in politics.

This may not be something China can easily do. Still, the transition in the objectives of Chinese economic policy goals—from an export-driven economy powered by foreign investment to an economy driven more by domestic demand and powered by innovation—is impressive. If the Chinese can succeed in making this change, they will avoid many of the problems that have hampered growth in the more mature economies of Europe and Japan.

Some Chinese complain that multinational high-tech firms have in effect "captured" Chinese researchers for their own use, weakening the national R&D base. This charge is not baseless: multinationals are careful to ensure that the benefits of the R&D in China that they fund accrue to them rather than to Chinese entities. This is no different, of course, than the case of an American researcher who works for a European company with facilities in the United States, but Chinese observers see this through a nationalist lens that makes it appear to be a disadvantage rather than a normal part of global commerce.

The participation of Chinese researchers in the efforts of multinational firms means we should also recognize that technology transfer from China to other economies is also increasing. Roughly 100 of the Fortune 500 companies have R&D centers in China. Firms from India, South Korea, and Japan are also building R&D centers—some Indian executives see these centers as a way to "capture" Chinese programming talent for Indian companies rather than for

---

[13] Kathy Chen and Jason Dean, "Low Costs, Plentiful Talent Make China a Global Magnet for R&D," *Wall Street Journal,* March 13, 2006; also, Organization for Economic Development (OECD), "China Will Become World's Second Highest Investor in R&D by End of 2006," April 12, 2006, at http://www.oecd.org/document/26/0,2340,en_2649_201185_37770522_1_1_1_1 ,00.html.

potential Chinese competitors. Ministry of Science and Technology officials noted that two thirds of "high-tech patents and inventions" in China were registered by foreign companies in 2003. More than 200 joint research and development centers have been established in China with leading foreign technology firms since 1990.[14]

Most new R&D facilities are concentrated in the telecom and software sectors and aim at the development of new products for the Chinese market, but some companies (such as Microsoft) are also focusing their centers on product development for the China or Asian markets. A National Science Foundation Report found that U.S. firms accounted for more than half of these, followed by Japan (26 centers), Germany (15), and the UK (14). In contrast, Korea opened only 3 and Taiwan none.[15]

Foreign companies estimate that the risk and cost of technological transfer is outweighed by economic returns. U.S. critics argue that there are externalities not considered by the firms in the form of costs to the larger U.S. national interest (e.g., national security, economic strength, technological leadership). These critiques usually fail to take into account two factors. First, there are clear benefits to the United States from trade with China—U.S. companies are more competitive globally, U.S. investors get higher returns, and U.S. consumers find lower prices. Second, the larger technological leveling among nations produced by globalization in the last decade means that efforts by one nation, no matter how large, to restrict China's access to technology will fail unless it has the support of all other industrialized nations. This support is noticeably lacking.

Western restrictions on technology transfer and export controls are now essentially irrelevant to the growth of China's IT industry. Despite Chinese complaints, they do not appear to have affected the pace or scope of growth. One semiconductor manufacturing equipment supplier said that it sells packages of equipment that provided everything needed to set up a production line. Only a few key pieces of equipment in these packages are controlled for export. The rest are uncontrolled. In selling a package to China, if they cannot ship controlled equipment from the United States, they would obtain it from manufacturers in Japan and ship the package in separate parts, some from the United States and some from Japan, rather than lose the entire sale to a foreign competitor.

Technology transfer restrictions may account for competitive advantage between foreign suppliers or investors—some U.S. firms may be at a disadvantage compared with their less-constrained European or Japanese

---

[14] The National Science Foundation found that 186 foreign companies opened research facilities between 1990 and 2001. U.S. firms accounted for more than half of these, followed by Japan (26 centers) and Germany (15) and the UK (14). In contrast, Korea opened only 3 and Taiwan none, consistent with an emphasis on production and assembly.

[15] U.S.-China Business Council, "Foreign Investment in China," http://www.uschina.org/statistics/ 2003foreigninvestment.html; Francisco Moris, "U.S.-China R&D Linkages: Direct Investment and Industrial Alliances in the 1990s," National Science Foundation, February 2004, NSF 04-306; "Pushing High-tech Key to Export Growth," *China Daily,* January 7, 2004, http://www.china.org .cn/english/scitech/84101.htm.

competitors, but at most, their effect on the IT capital goods market in China may be to allow European or Japanese firms to extract higher prices from Chinese customers. The explanation for this irrelevance lies in the end of the Cold War and with it the end of Cold War controls on technology transfer. European nations (particularly Germany) and Japan had lifted restrictions on transfers to China of advanced industrial equipment by the mid-1990s.

U.S. export policy restricted transfers of semiconductor manufacturing equipment to China by U.S. firms to two or three generations behind state-of-the-art, but other countries have not been similarly constrained. This is particularly true of Taiwanese firms, despite Taipei's announced policy to block the transfer of advanced manufacturing equipment to China. Taiwanese companies and investors are the leading foreign developers of China's microelectronics industry. This has prompted fears among some Taiwanese officials that Taiwan risks being "hollowed out" by the shift of manufacturing to China. These fears now seem exaggerated. All other major suppliers of semiconductor manufacturing equipment—the Netherlands, Germany, and Japan—have told the United States that they will not block sales to China. European and Japanese officials have questioned what contribution semiconductor manufacturing equipment has made to military capabilities and proliferation and whether controlling this equipment remains strategically relevant.

Although there was a consensus in the 1980s between the United States and its allies to control technology transfers to the Soviet Union, this consensus no longer exists. The clearest indicator of the lack of consensus lies in the demise of COCOM, an organization linked to NATO that controlled sensitive exports to the "communist bloc." By 1991, COCOM was moribund, and U.S. allies demanded that it be disbanded, effectively ending multilateral cooperation on most commercial exports. Europe's concerns in Asia are primarily commercial, making it very hard to win support for security-based export restrictions on most goods, including semiconductors and semiconductor manufacturing equipment.

The successor regime to COCOM (known as the Wassenaar Arrangement) is ineffective. The result is that U.S. regulations for the export of semiconductors and semiconductor manufacturing equipment, which are still largely based on COCOM, are increasingly unilateral and outdated. Some in the United States continue to call for an effort to restore multilateral controls on dual-use technologies, but this does not fit current economic and political trends—France, Russia, and Germany blocked a strong replacement to COCOM in the 1990s and are unlikely to have become more sympathetic in the interval. The Netherlands and Japan, whose firms are key players in the semiconductor industry, do not accept a need for restriction.

This divergence on technology transfer became apparent as early as 1996, when China launched Project 909, a key element of the 1990 Ninth Five-Year Plan, to develop a national microelectronics industry. Project 909 involves joint ventures between Chinese and foreign firms. China has invested $1.2 billion in the 909 project, according to press reports.

When Project 909 was announced, China looked to companies in the United States and Japan for foreign partners. The United States began a lengthy internal review to consider whether to permit exports of semiconductor manufacturing technology. But shortly after the United States began its review, Japan announced that it had approved the participation of a major Japanese semiconductor manufacturer in 909 and the transfer of advanced semiconductor manufacturing technology. The transfer was covered by a short written agreement between China's premier and Japan's prime minister in which China promised not to use the semiconductor manufacturing equipment for military or proliferation purposes. The United States sought to discourage the transfer, but the Japanese responded that they did not see the connection between manufacturing semiconductors and improved military capabilities. Japan's decision to go ahead with Project 909 opened the door for other companies to begin transfers of advanced semiconductor technology to China.[16]

China's WTO commitments will have a greater effect in reducing technology transfer than Western restrictions. Most observers say that China is making a good faith effort to implement its WTO commitments. WTO compliance is reducing the number of joint ventures and the amount of technology transfer as China brings its investment and IP rules into line with international practice. WTO rules, for example, forbid technology transfer requirements as a condition of approval for foreign investment. WTO accession may also reduce technology transfer because of the emphasis it has given to China's efforts to develop indigenous standards for IT products. China continues to show interest in using standards and regulatory policies to create advantages for its domestic IT industry.

The recent effort with WAPI (Wireless LAN Authentication and Privacy Infrastructure) showed the combination of security, commercial advantage, and national prestige that motivate China to pursue its own standards. WAPI has been portrayed by China as an indigenous solution to the problem of securing wireless communication. Privately, most foreign firms viewed the regulations as an ill-disguised effort to increase market share for Chinese companies and to gain access to Western intellectual property. Multinational company representatives in China reported that smaller Chinese companies saw WAPI as an important commercial opportunity. China may have preferred WAPI, however, for security reasons.

When China first unveiled WAPI in 2003, it announced that foreign manufacturers had six months to begin to use it in their products. Since the standard was classified, China did not release it to foreign companies, but required them to partner with 24 Chinese institutions or companies to bring wireless products into compliance with the new rules. Implementation of the new standard (probably in the form of a module that would be added to the Western product) could be carried only by a designated Chinese company. The encryption algorithm used by WAPI was classified as secret until January 2006 and has still

---

[16] The author was one of the negotiators in the ill-fated U.S. effort to persuade the Japanese to curtail their involvement in Project 909.

not been widely disseminated. This lack of dissemination and review worked against WAPI's adoption as an international standard.

The 2003 announcement met with resistance from Western companies and governments, particularly from the United States. Chinese officials noted, somewhat defensively, that "among the world's 16,000 international standards 99.8 percent are made by foreign institutions." China has 19,278 national standards for a broad range of products, and standard setting was a key management tool for the Soviet-style centrally planned and directed economy found in China until the 1990s. Coming from this background, the Chinese argue that the lack of a national standard "will make it hard for the government to regulate the [WLAN] market" and that a failure to adopt a secure WLAN standard will jeopardize the security of Chinese networks.[17] However, China agreed to indefinitely postpone the requirement to use WAPI and instead sought to have it adopted as an international standard.

However, WAPI was rejected by a vote in the International Standards Organization (ISO) in May of 2006. Chinese officials were displeased and charged that American Institute of Electrical and Electronics Engineers and the U.S. government had organized a campaign to block adoption of the new standard. This says as much about how the Chinese view of government's role in standard setting as it does about WAPI—China assumed that governments played a central role, while in the United States this was seen as essentially a private sector decision. WAPI opponents at the ISO repeated earlier charges that the Chinese standard was not an improvement and functions as an illicit mechanism for forced technology transfer.

WAPI will not be China's last foray into standard setting. China is looking at domestic standards for Radio Frequency Identification tags (RFID), 3G telephony, video compression, and DVDs. American IT companies say that standards will be the trade battleground of the future with China. However, many industry representatives believe that China has two goals, in that even if it does not succeed in imposing domestic standards, it will gain additional leverage in negotiating with international standards bodies. WAPI and the emphasis on the use of indigenous standards and products highlight a particular tension between Beijing's desire for control and the openness required to be globally competitive.[18]

The standards process, if used astutely, can create a barrier to trade while avoiding potential WTO complications that could arise from the use of other

---

[17] *People's Daily,* cited by China Education and Research Network, "China, Wi-Fi Fight for WLAN Standard," http://www.cernet.edu.cn/20040205/3098441.shtml.

[18] China National [Standards] Body, "Unjust Activity, Undue Process, Unfair Results: Ethical and Procedural Violations in WAPI-11i Fast-track Process," April 21, 2006, http://www.chinabwips .org/doc/Unjust%20Activity,%20Undue%20Process,%20Unfair%20Results.pdf; *People's Daily,* "Chinese WAPI Delegation Calls for Diplomatic Support after "Unfair Treatment," June 9, 2006, http://english.people.com.cn/200606/09/eng20060609_272428.html; CIO Magazine, "China's WAPI Not Going Down without Fight," May 3, 2006, http://www.cio.com/blog _view.html?CID=21560.

kinds of barriers to trade. The Chinese also see development of domestic standards as a means to overcome licensing difficulties and costs it faces in the use of foreign intellectual property. Use of the standards process to gain competitive advantage is a normal business practice. American and Japanese companies expect the Chinese to become very influential negotiators in the standards process. One electronics company currently involved in a standards dispute with China said that his company feared retaliation in other areas because of the standards dispute and was looking for ways to placate China, such as investing more funds in Chinese R&D centers and sending senior executives from Japan to high-profile events in China.[19]

Semiconductors

"Our microchips are all imported. We do not make our own chips. Our whole industry is built on sand. These chips are like the heart of a person. If the source of supply is cut off, our heart will stop beating."

Wu Dexiang, member, National People's Congress; director, China Microelectronics Centre Study Committee

China's national policies emphasize the creation of a domestic semiconductor industry for both economic and security reasons. The emphasis appeared as early as the "Four Modernizations" development program of the 1970s. Prior to the 1990s, China had a microelectronics industry before it opened its economy to foreign investment, but it was based on outmoded Soviet technology, was inefficient, and lagged several generations behind Western firms in the products it churned out. This is no longer the case.

Two broad factors explain the rapid change in China's semiconductor production. First, the Chinese government created incentives for foreign participation in domestic semiconductor production. Second, Taiwanese investors and semiconductors executives, seeing an immense opportunity for savings and rapid growth, moved manufacturing operations to China to take advantage of cost differentials and subsidies. Other foreign semiconductor manufacturers, from Japan and the United States, rapidly followed suit, but it is the Taiwanese influx that more than anything else explains the rapid growth in China's microelectronics industries, including semiconductors.

In light of the major role played by Taiwanese and other foreign semiconductor manufacturers in building production capabilities in China, it is fair to ask if this is a Chinese semiconductor industry or the semiconductor industry in China. The major semiconductor manufacturers locate their production centers in many different countries. What is different here is the effort by the Chinese government to encourage the development of national champions in chip production—

---

[19] Zhong Jing, "Comments on China's Enterprises Patent Strategy," *China Economic Net*, June 9, 2004, http://en.ce.cn/Insight/t20040609_1034885.shtml.

Chinese-owned companies producing chips from Chinese designs. The intent is to challenge and to gain national leadership, but intent alone does not guarantee success.

As of mid-2005, there were approximately 200 assembly and test companies in China, along with 40 domestic manufacturers of equipment for the semiconductor and related microelectronic industries. There is a strong multinational presence in China's chip industry—8 of the 10 largest semiconductor manufacturing facilities, or "fabs," are owned or operated in partnership with Taiwanese, U.S., Japanese, or European companies. There is some dispute over the number of fabs planned or being built because fabs are seen as prestigious, and many local governments are eager to acquire (or at least announce) a fab for their jurisdiction.[20]

Four of these fabs (all with significant foreign ownership) will use current generation production technologies in wafer size or feature size according to press reports and remarks by company officials. Taiwanese and Japanese producers report that their Chinese facilities are operating at the 0.35 or 0.25 micron level (a micron is a measure of feature size; chips with smaller feature sizes are faster and more powerful) and are moving to 0.18 microns. For all practical purposes fabs located in China are on, or close to, the cutting edge of semiconductor technology.

China's experience is not unique, and to some degree the policies it adopted to promote a domestic semiconductor industry appear to be derived from the experience of other Asian nations. China is the latest in a line of successful Asian entrants into semiconductors. In the 1980s, the semiconductor industry was reshaped by the entry of Japan into the semiconductor manufacturing sector and in the 1990s by the entry of Taiwan and Korea. The current cycle is being driven by the entry of China. According to the chairman of a leading Taiwanese chip manufacturer, "The center of gravity of the semiconductor world is slowly but inexorably shifting to China." American and Japanese companies take the view that China (or companies and fabs in China) are becoming part of an interdependent semiconductor manufacturing network centered on the Pacific Rim.[21]

Semiconductors vary widely according to their function. Chips for the central processing unit or CPU (the "brain" of the modern computer) are among the most complex to design and manufacture. U.S. companies lead the CPU market. Korean, Taiwanese, and Japanese firms lead the market in the memory chip, another crucial component for computers. Telecommunications uses a range of specially designed chips, and video gaming requires specialized chips. Automobiles and other consumer applications use microprocessors. The complexity and difference involved in manufacturing these different varieties, along with the cost, mean that for high-end production a few large firms dominate the market. Established U.S., Japanese, Korean, and Taiwanese companies are the

---

[20] A "fab" is a facility for the fabrication of semiconductors. Mark LaPedus, "20 New Fabs to be Built in China," *EE Times,* June 9, 2005, http://www.eetimes.com/news/semi/showArticle .jhtml?articleID=164301951.
[21] Morris Chang, "Remarks to the US-Taiwan Business Council," December 2004.

market leaders. The top 20 chip manufacturers, none of whom are Chinese, produce two thirds of the world supply of chips.[22]

The market for chips itself is highly competitive, margins on chips can be small, and microprocessors are increasingly a "commodity" product, albeit a commodity that rests on advanced and expensive scientific and manufacturing capabilities. A fab costs between $3 billion and $4 billion to build and will be outmoded after five or six years. These costs make semiconductors a difficult industry to enter. But semiconductor manufacturing is being reshaped by disaggregation in production. Instead of one company doing everything from taking in silicon to putting tested chips in boxes, the manufacturing process is being divided among different specialized firms that provide design, wafer production, foundries, assembly, and testing. This "fabless" approach to semiconductor manufacturing now accounts for about 17 percent of the semiconductor industry and continues to grow.[23] The global trend toward disaggregation has made it easier for Chinese firms to find a place in the production cycle of the semiconductor industry.

The Tenth Five-Year Plan called for 25 new fabs to be constructed, and a 2000 State Council Directive called for an investment of $10 billion in semiconductors. Western industry experts disagree on the number of fabs projected to be built in China in the next few years. Equipment manufacturers (who would sell to the new fabs) put the number at 20 while semiconductor manufacturers (who would compete with the new fabs) project only 10 to 15 new fabs.[24] China has invested heavily in a few semiconductor companies and offered a range of incentives for foreign investment in the semiconductor industry.[25] However, China's efforts to design and develop its own semiconductors have had mixed results. Production of sophisticated kinds of chips eludes Chinese entities. Although the Chinese goal is to end dependence on foreign technology, the "Dragon chip," a well-publicized national effort to create a CPU, was developed with help from IBM and Transmeta. Its runs on a Chinese version of Linux and operates at a speed of 266 megahertz (MHz) that, if accurate, would make this chip equivalent to a Pentium II.

[22] Bolaji Ojo, "The Global Top 20," Electronics Supply and Manufacturing, July 1, 2005, http://www.my-esm.com/showArticle.jhtml?articleID=165600157.
[23] Fabless manufacturing involves the customer's designing the components but then contracting the actual manufacture to another entity rather than acquiring its own manufacturing capabilities.
[24] Mark LePedus, "SEMI, SIA Spar over Projected Fabs in China," EE Times, July 12, 2005, http://www.eetimes.com/showArticle.jhtml;jsessionid=W42HPBX2P2LB2QSNDBGCKH0CJUM EKJVN?articleID=165701697.
[25] Semiconductor Industry Association data; Department of Commerce, "Year-End 2003 Summary of Developments in China's Semiconductor Industry," March 2004; Michael Pecht, Weifeng Liu, and David Hodges, "Trends in China's Semiconductor Industry," Semiconductor International, September 1, 2000, http://www.e-insite.net/semiconductor/index .asp?layout=article&articleid=CA163413; "IC Demand on the Rise in China," Electronic News, March 24, 2004, http://www.ic-china.org/ENICCHINA/jingying/page_0000004.htm.

Dragon also faces a new challenge as a result of AMD's decision to license its x86 technology to China. Public accounts of the arrangement suggest that AMD will get licensing fees from Chinese companies that choose to use their design. AMD will also get access to derivative technology. But the transfer creates another, and proven, alternative to the Dragon chip for Chinese companies.[26] They can now buy chips from foreign sources, license AMD technology, or, in what may be the least attractive option, use the Dragon.

Although much of the impetus for the Dragon chip came from PLA concerns about relying on U.S. semiconductors, China's military would be at a disadvantage if it used the Dragon chip to replace Western semiconductors. The first generation Dragon was comparable to an Intel chip from 1995 and operates at between 200 and 260 MHz.[27] The chip is itself not commercially competitive, and China would have to use nonmarket incentives or restrictions to get people to prefer Dragon to other chips. The hope may be that by continuing to invest in chip design, by gaining experience, and by working with foreign partners, China will eventually develop an indigenous CPU that meets world standards; for now, however, Dragon is an expensive prestige project.

China's efforts to develop an indigenous semiconductor industry suffered an embarrassing blow. A researcher at Jiaotong University in Shanghai (who had trained and worked in the United States) announced that he had developed the "Hanxin" chip for digital signal processing (DSP chips can be used in a range of consumer and military goods, including cameras and mobile phones). The scientist also started his own company, also called Hanxin, to make the chips. Designing chips is a complex task, and turning those designs into actual chips even more complex. Hanxin won national acclaim until it was discovered that the chip did not exist. Early prototypes were actually chips made by an American company where workers had removed the U.S company name and replaced it with Hanxin. The story illustrates both the risk of IP theft and the difficulties of chip manufacturing; it is not clear that Hanxin could have successfully produced a copy of the American chip.[28]

The most successful semiconductor producers in China are either joint ventures companies or foreign-owned. The most advanced "Chinese" semiconductor manufacturing facility is SMIC (Semiconductor Manufacturing International Corporation)—China's largest chip manufacturer. It is Taiwanese-owned, and its chairman Richard Ru Gin Chang lived for several decades in the United States where he worked for Texas Instruments. SMIC currently has three 8-inch-wafer

---

[26] Wu Zhong, "Driving out the Dragons," *The Standard,* November 7, 2005, at http://www.thestandard.com.hk/news_detail.asp?pp_cat=15&art_id=5065&sid=5319034&con_type=1.

[27] A measure of chip performance. Current CPUs on the market operate at 20 times this speed. The Dragon II, released in April 2005, is 3 to 4 times faster than the Dragon I, according to Chinese sources, but faces charges that it infringes heavily on MIPS technology developed elsewhere.

[28] BBC, "Fake Chip Research Shocks China," May 15, 2006, http://news.bbc.co.uk/1/hi/business/4771583.stm; David Barbosa, "In a Scientist's Fall, China Feels Robbed of Glory," *New York Times,* May 15, 2006.

fabs in Shanghai that will increase capacity by 70 percent this year. SMIC recently purchased Motorola's Tianjin 8-inch fab. SMIC also planned to open a 12-inch fab in Beijing late in 2004 and two more 12-inch fabs in 2005 and 2006. SMIC received long-term leases at concessionary rates, training subsidies, and other incentives from the Shanghai government to locate its fab there. SMIC is a good example of the semiconductor industry in China: although located in China, its CEO is Taiwanese; its management and R&D staff are a blend of European, U.S., and Japanese citizens; and its customer base is international.

Vimicro, another leading Chinese chip manufacturer, is among the top 20 global producers of multimedia microprocessors for camera phones and other consumer devices. Vimicro was founded in 1999 by four Chinese entrepreneurs from Northern California—two graduated from Berkeley and one from Stanford. It was supported by the Beijing municipal government and the Ministry of Information Industries, and its chairman, John Deng, was named Best Overseas Returnee Entrepreneur by China's president Hu Jintao and prime minister Wen Jiabao in 2003. Listed on the NASDAQ in November 2005, Vimicro has partnerships with leading Western IT firms, including Microsoft. Its revenues of $14 million in 2004 do not approach the top IC firms with revenues in the billions, but it has a leading share in some multimedia applications and is the darling of the Chinese government.

U.S. government figures show that semiconductor production in China has increased at a rate of 25 percent a year since 2000.[29] One estimate suggests that China produced more than 10 billion semiconductors in 2003 worth $8.3 billion—about 4 percent of world production.[30] But this is only about 20 percent of the semiconductors needed to meet domestic demand in China. The rest must be imported. China became the largest market for chips of all kinds in 2005. The market for semiconductors in China is expected to continue to grow at 30 percent a year, driven mainly by demand for products like cell phones, digital cameras, automobiles, and other consumer products. Beijing hopes to raise the share of domestic chip production to more than 50 percent, but this is a long-term goal that may not be met by Chinese-owned companies.

Some Taiwanese semiconductor industry leaders expect China to follow Taiwan's progress in moving up the value chain in semiconductor manufacturing, possibly at an accelerated pace. Thirty years ago, U.S. companies opened assembly plants in Taiwan. Twenty years ago, the first foundries appeared in Taiwan. A decade ago, Taiwanese companies began to open design centers. The United States still leads in design and in determining chip architecture, but Taiwan has moved into second place, and Taiwanese firms plan to move into more IP-intensive aspects of the semiconductor manufacturing process.

Beijing also hopes that it can make China a center for IC design. This will be a difficult task in an already competitive global market. Design is one of the most

[29] U.S. Commercial Service, "China: Computers and Telecommunications," http://www.buyusa.gov/china/en/computers.html.
[30] PriceWaterhouseCoopers, "China's Impact on the Semiconductor Industry," April 2005.

complex and valuable activities in microprocessor production. It requires experience, ownership of (or access to) a substantial amount of intellectual property, a highly skilled workforce, and a close connection to scientific research. Currently, the United States leads in chip design, followed by Taiwan. China has almost 500 design companies, but these companies generated only $1.9 billion in revenue (in a global market worth $33 billion). The China Semiconductor Industry Association estimates that perhaps only 20 of these firms are competitive.[31]

Semiconductor industry executives attribute the problems of China's chip designers to weaknesses in understanding the market and using silicon to satisfy market applications. China has several thousand chip designers, but most lack the experience needed for chip design. High-end applications, such as CPUs, are beyond their reach. The industry view is that competence in design requires different, more advanced skills that China does not yet possess. Despite this, some industry analysts expect that locally designed and manufactured products will dominate some sectors of the China market, such as chips and ASICs (specialized chips designed for a specific application) for telecommunications and consumer entertainment products.

The effect of China's government-subsidized semiconductor manufacturing and design industries will be to squeeze smaller semiconductor producers in other countries. Overall, the main beneficiaries will be the U.S. and Asian firms that currently dominate semiconductor manufacturing, as they incorporate China into their production chains and take advantage of China's huge market. China is still heavily dependent on foreign expertise and investment for semiconductor production, particularly from Taiwan and the United States. Imports by China of both chips and technology will be the norm for many years to come. The result is as much increased interdependency as competition.[32]

Fears that China would come to dominate semiconductor production now seem overstated for several reasons. Much of the production is absorbed by the domestic market. The most successful firms are multinational or foreign-owned. Although some analysts believe that manufacturing costs for fabs located in China are 10 percent to 30 percent lower than their foreign competitors,[33] foreign firms have been skillful in capturing these benefits, reducing any competitive advantage indigenous producers might have gained. Foreign semiconductor technology and manufacturing equipment are still essential to China. The most likely outcome is that in the future, China will duplicate the experience of Japan, Taiwan, and Korea, with one or two Chinese firms joining the ranks of international competitors in semiconductor manufacturing while the market remains dominated by a few large multinationals.

---

[31] Amanda Liang and Rodney Chan, "A Look at China's IC Design Houses," *Digitimes,* September 16, 2005, http://www.finanznachrichten.de/nachrichten-medien/digitimes-6.asp.
[32] Applied Material Corporation, "China and Global Semiconductor Industry: Update and Outlook," May 27, 2004.
[33] Andrew Lu, Citigroup Smith Barney, and Rick Hsu, Nomura Securities.

Software

"As a big country, China could not be satisfied with the status of always being low-end software users or developers."

Li Quqiang, Ministry of Science and Technology

China has also identified the creation of an indigenous software industry as a national priority. It hopes that a combination of government support, partnership with Western firms, local talent, and a greater use of open source software will let it repeat its successes in IT assembly and manufacturing. China wants to build a globally competitive industry for its economic benefits, but also to reduce the risks it perceives come with the use of foreign software.

Achieving this goal may take years. China is entering a highly competitive and complex global industry in which there are already many players with a variety of motives when it comes to China. Some foreign firms work with Chinese software companies in the hopes that assisting in the development of Linux-based operating systems or applications will undercut their global competitors. Japan and Korea hope to collaborate with China to create a northeast Asian software industry that can challenge the American advantage in software. Indian software executives fear China as a potential competitor, but also hope to take advantage of China's market and programmers to accelerate their move to global status.

China's domestic software market continues to grow rapidly, driven by Internet use and the acquisition of personal computers. Business sources say that software sales in China reached $6.12 billion in 2003.[34] Some analysts predict this will grow to $10 billion by 2007. Chinese government figures show the domestic software market grew from $2.5 billion in 1996 to $11.9 billion in 2002.

Industry sources estimate that indigenously developed software currently accounts for only a third of the Chinese market. Western software companies report that demand for software in China is primarily for low-end application software and games. Indigenous efforts to develop commercial software concentrated first on interface applications and on Chinese-language word processing. Government agencies, banks, and telecommunications firms have been the largest source of demand for more advanced products. Most Chinese firms still have only limited demand for high-end business software (such as Customer Relations Management software) as many firms do not have complex internal processes, nor do they yet face the same pressure found outside China to lower costs through the use of technology. American software companies expect that this will change over time as China's businesses become more sophisticated, but for now overall levels of IT spending by companies in China remain relatively low. This means that Chinese software firms, if they hope to sell the high-end products and services that are more profitable, will need to sell internationally in a

---

[34] People's Republic of China, Ministry of Information Industry, Center for Computer and Microelectronics Industry Development, http://www.ilog.com/corporate/connection/summer_2002/CompanyNews.htm.

much more competitive global market where they are at a relative disadvantage compared to foreign software producers.[35]

A survey by the Hong Kong Trade Development Council in November 2000 found more than 2,000 Chinese firms engaged in the development, production, sales, maintenance, or servicing of software. Another 3,000 firms were involved with software as a secondary line of business (especially in sales and services). Perhaps 1,000 firms are devoted to software development, but the majority of Chinese software firms employ fewer than 100 persons. In 2004 11,000 "companies engaged in software and information service," 6,000 of which have "exclusive IP rights and R&D forces" and "6,000 certified software firms."[36] Indian and American software executives believe that the small size limits the scope of software projects that Chinese firms can undertake and that Chinese firms lack key software integration skills.

State-owned enterprises make up 30 percent of Chinese software firms. Privately or collectively owned firms make up 40 percent. Foreign-owned and joint ventures make up 30 percent of the software companies, but this figure does not reflect the large number of Chinese firms that have partnership arrangements with foreign companies.[37]

As with semiconductors, building a national software industry is a goal for the 863 Program. The program identifies software as one of its "key projects." The Ministry of Science and Technology's Torch Program is also important for this national effort. It funds the China Offshore Software Engineering Project (COSEP), which supports software parks where companies will write code for the American market. COSEP will invite foreign outsourcing and project management service experts to join the program and will eventually help Chinese companies establish offices overseas to enter the global software market. Although NASSCOM's assessment is that China's software industry, despite Torch and other efforts, will not pose a challenge to Indian firms in the outsourcing market for the foreseeable future, there is some nervousness among Indian software companies.[38]

---

[35] Matei Milhalca, "China: Rising Software Power?" June 14, 2004, *Rediff.com,* http://www.rediff.com/money/2004/jun/14spec.htm; Chang Tianle, "Leading Software Firms Targeting Overseas Market," *China Business Week,* June 11, 2004, http://www.chinadaily.com.cn/english/doc/2004-06/11/content_338698.htm; U.S. Department of Commerce, International Trade Administration, "Telecommunications and Information Technology Market Opportunities for Small and Medium-sized Enterprises: Export IT China," April 2003; U.S. Commercial Service, "Software Industry 2003," February 2004; Li Wuquiang, "Open Source Software and China's Software Development," presentation on May 20, 2004, at the conference, "Strategies for Building Software Industries in Developing Countries," University of Hawaii at M noa.
[36] See *tdctrade.com,* http://www.tdctrade.com/report/mkt/mkt_040402.htm.
[37] "Business Alert—China," *tdctrade.com,* http://www.tdctrade.com/alert/chwto0011ai.htm; "China Threatens Indian Eminence," *Wired,* http://www.wired.com/news/business/0,1367 ,41656,00.html?tw=wn_story_related, February 23, 2001.
[38] See http://www.most.gov.cn/English/Programs/torch/menu.htm.

A project named "Jade Bird" exemplifies government support. A long-standing government-funded program to increase China's software capabilities, Jade Bird began in the 1980s, involves 20 institutions and 300 researchers, and is led by Yang Fuqing, a dean at Beijing University. Yang has set up a software company (also called Jade Bird) that commercializes research from the program. Jade Bird, which focuses on large-scale industrial software development, is building a "national component library" for software production.[39]

Beijing and Chinese companies pin many of their hopes for a globally competitive software industry on the use of open source software, particularly the Linux operating system. Linux and open source software are an important source of technology for China's IT industry. Access to the Linux kernel is crucial for the development of an indigenous Chinese operating system. China's leaders hope that open source software will allow China's software industry to move rapidly into the production of products that, if not globally competitive, will at least dominate the domestic market. They believe that open source products are more secure. Finally, a use of open source software avoids the licensing (or piracy) issues raised by the use of proprietary intellectual property. The Chinese are particularly concerned about the disadvantages and costs of being dependent on foreign IP and are eager to create national alternatives. Surveys show that nearly half of Chinese programmers use Linux and that nearly three quarters say that they plan to write applications for Linux in the future.[40]

By itself, access to open source software is not enough to ensure the development of viable Chinese software industry. Other governments besides China share the hope that they can create national software industries based on open source. Access to open source software clearly lowers the cost of entry into the industry. However, this in itself is not sufficient to be competitive. Access to code provides only part of what is needed for a successful software company. In particular, many nations underestimate the necessity of having an effective legal structure (especially sufficient protections for intellectual property) and a population with the necessary marketing and management skills if they are to compete in the global software market. For China, the most valuable element provided by open source software beyond a transfer of technology may be its role as a vehicle for partnership with more experienced foreign firms.

U.S. and Indian software company representatives say that though the skill of individual Chinese programmers is high, Chinese firms lack the management and integration skills needed to succeed with large-scale software projects like an operating system. India benefits from a greater number of returnees with experience in large U.S. software companies. NASSCOM executives cited one measure of software capability—the Carnegie Mellon Software Institute's Capability Maturity Model (CMM)—where in late 2001 India had more than 30 firms capable of work at the highest level while China had only one. For China,

---

[39] Beijing University, Software Engineering Institute, "Brief Introduction of Professor Yang Fuqing," http://www.sei.pku.edu.cn/en/yangfuqing.jsp, and "An Introduction to Jade Bird Project," http://www.sei.pku.edu.cn/en/jadebird.jsp.

[40] U.S. Commercial Service, "China: Semiconductors and Software," 2004.

the most valuable element provided by open source software beyond a transfer of technology may be its role as a vehicle for partnership with more experienced foreign firms.[41]

CMM is based on quality standards defined by the Software Engineering Institute (SEI). SEI is a federally funded research and development center sponsored by the U.S. Department of Defense and operated by Carnegie Mellon University. CMM ensures that various software processes are clearly defined to ensure quality control and delivery. Its scope covers the software development process, defining project scope, time, software design, development, and testing. CMM provides a standard for various software production and project management processes that allows companies to assess the risks and benefits of outsourcing software development. The highest CMM level—level 5—means that a company can ensure predictable, consistent results when it undertakes a software project. Of the 70 level 5 companies in the world, 50 are in India.

Although more than 100 companies in China have undertaken and achieved CMM/CMMI certification at various levels, most of them (more than 80 percent) are at levels 2 or 3. Chinese companies are progressively moving toward higher CMM levels, but only one is at level 5.[42]

Continued government support means that Chinese software companies will improve their CMM scores. Information security goals help explain this continued support. Beijing's long-term plans for more secure national networks depend on success in an effort to move to domestically produced software. In part, this is explained by China's suspicions of Western (particularly U.S.) software. Chinese officials suspect that this software has been modified to allow foreign entities access to Chinese computer networks. These suspicions have no basis in fact, but despite this, they continue to dog Western software companies and explain some of the effort to encourage indigenous operating systems like Red Flag and other locally made, open source–based software.[43]

Chinese officials have denied Chinese press reports that government policy banned ministries from using Microsoft Windows,[44] but there are reports that the State Council imposed a requirement in November 2003 that ministries must purchase indigenously produced software in the next acquisition cycle. Despite this edict, most users in and out of government seem to prefer pirated versions of

---

[41] Gartner Group, "India and China: The Race to Capture Software Business," July 2002; Ted Tschang, "China's Software Industry and Its Implications for India," http://www.oecd .org/dataoecd/30/31/2401094.doc; Kate Walsh, "Foreign High-Tech R&D in China: Risks, Rewards, and Implications for US-China Relations," Henry Stimson Center, 2002.

[42] "Outsourcing in China," *Insidernews,* October 3, 2004, http://sherman.linuxhall .org/?q=node/view/216.

[43] In Roman Loyola, "China: The Republic of Linux," August 13, 2002, Zhao Xiao-Liang, senior vice president of Red Flag Software, discusses China's reaction to the "NSA key" and security concerns about U.S. software; see http://www.techtv.com/screensavers/ linux/story/0,24330,3395670,00.html.

[44] *People's Daily,* "China Approves Microsoft Windows 2000," October 15, 2003, http://english.peopledaily.com.cn/200310/15/eng20031015_126076.shtml.

Western software and ignore commands to stop using it. American companies suspect that the pattern of issuing directives to purchase local products and then retracting them when they attract attention could indicate an internal debate between security and economic agencies on the benefits of moving toward reliance on indigenously produced software products.

China has been particularly concerned with the security of operating systems. U.S.-made operating systems run on perhaps 90 percent of the country's computers (albeit in pirated versions). He Dequan, an influential academician with connections to the Ministry of State Security,[45] wrote in a 1999 *People Daily*'s article on cyber security: "Over the long term, China needs to develop its own operating system. The present trend towards free software provides a good opportunity." He noted that "the United States leads the tide towards globalization and is using information hegemony to dominate the world."[46]

Several U.S. companies have allowed Chinese officials access to their code in an effort to assuage these concerns, and China's extensive government review and certification process for software, which involves multiple agencies with overlapping jurisdiction, can give Chinese agencies and firms additional insight into foreign code. One company, Microsoft, responded to Chinese concerns about intentional vulnerabilities by creating the Government Security Program (GSP), which gives governments (including China) a chance to review the source code for Windows to satisfy their security concerns.

The review of GSP was intrusive, but it is only part of a larger and more intrusive regulatory approach. China has 16 published regulations that relate to Internet security, and they are reinforced by numerous administrative directives. These regulations directly affect the development of IT products, particularly software.

Microsoft says that it created GSP "to address the unique security requirements of governments." Microsoft had long shielded its Windows source code from scrutiny; the decision to allow Chinese government agencies to review it, despite the risks of intellectual property losses, shows both the depth of the Chinese government's concern and the interest of U.S. firms in the China market.[47] The transparency provided by the GSP should effectively allay Chinese security concerns, but China is likely to continue to pursue efforts to develop its own operating system for reasons of commerce and prestige.

---

[45] He Dequan is the director of the Science and Technology Commission of the Ministry of State Security and a frequent speaker and writer on China's cyber security situation. In 2001, he briefed the State Council, including Premier Zhu Rongji and Vice Premier Li Lanqing, on the need for greater government attention to information security.

[46] He Dequan, "We Must Act Immediately to Protect Our Information Security (Information Technologies Forum)," *People's Daily,* October 17, 1999, p. 4, at http://www.usembassy-china.org.cn/sandt/pdinfosec.html. Dequan is a frequent speaker and writer on China's cyber security situation. In 2001 Dequan briefed the State Council, including Premier Zhu Rongji and Vice Premier Li Lanqing, on the need for greater government attention to information security.

[47] *CNet Asia*, "China to See Windows Code," February 28, 2003, http://asia.cnet.com/newstech/systems/0,39001153,39116601,00.htm.

Although GSP will extend the marketability of U.S. software in the China market, open source software remains very attractive to the Chinese for several reasons. China's leaders clearly hope that access to open source software will provide an infusion of technology that will allow China's software industry to move rapidly into the production of globally competitive products. Many Chinese share the belief that open source products are more secure, and one Chinese official noted that "all governments, including the Chinese, have concerns about security. Even the American government doesn't want foreign parties to know their secrets. With Microsoft, you don't have the source code, although you can check it. With Linux, it's open and you can thoroughly study it."[48]

Security concerns led to a government-sponsored development effort in the 1990s to develop a Chinese operating system. Known as Red Flag, the operating system did not work very well. The Software Research Institute of the Chinese Academy of Sciences (CAS) developed Red Flag based on the Linux open source operating system. Although access to the Linux kernel provided a crucial transfer of technology, Western experts who worked with the product say that the initial variants of Red Flag produced by the Software Research Institute were unstable and, unsurprisingly, had application compatibility problems.[49]

As a government software program, Red Flag was not a success. However, the fortunes of Red Flag improved as it became more commercial and found Western assistance. In 2000, CAS spun off Red Flag, helped by funding from the Ministry of Information Industries and from an investment arm of the Shanghai Municipal government.[50] The new, privatized Red Flag quickly entered into agreements with a large number of leading American IT companies. In the case of Red Flag, for example, one key partnership is with Miracle, a Japanese Linux company that is 58 percent–owned by American software giant Oracle. Other examples of Chinese operating systems where international collaboration was important in the development process include Cosix, a variant of the UNIX operating system developed jointly with Compaq Corporation, and Yangfan Linux, developed with help from American programmers and software engineers from Taiwanese government research organizations.[51]

---

[48] Li Wuqiang, deputy director, Ministry of Science and Technology, High Technology and Industrialization Department, quoted in the *Los Angeles Times,* "Linux Seen as Saviour from Microsoft's Grip," August 13, 2004, http://www.ebusinessforum.com/index.asp?doc_id=7126&layout=rich_story.

[49] "Red Flag Flies at Gates," *Computer Business Review*, January 11, 2000, http://www.cbronline.com/print_friendly/be7cac11873d764080256d350047d2ae.

[50] Funding came from the venture capital firm New Margin. New Margin was created by Shanghai Alliance Investment Company, an investment vehicle for the Shanghai Municipal Government, and the China Foundation of Science and Technology for Development, an investment vehicle for the State Development Planning Commission, State Economic and Trade Commission, and Chinese Academy of Science.

[51] "Chinese Government Raises Linux Sail," *Linuxworld,* August 2002, http://www.computerweekly.com/Article114972.htm; "Chinese Developer Delegation Visits to the TRON Association," http://tronweb.super-nova.co.jp/tronnews98-11.html.

Red Flag has helpful official connections (for example, Jiang Mianheng, the son of former Chinese president Jiang Zemin, is an executive at Red Flag), and it has been helped by government procurements and awards (named Most Trusted Brand of Operating System in 2003, for example). Red Flag claims 5 to 10 percent of the market for operating systems in China and hopes to expand this to 25 percent. All Linux-based operating systems, including Red Flag, are growing in market share, but are estimated to account for only 15 percent of operating systems in China at most.

The fortunes of the Red Flag operating system mirror the experience of other Chinese government IT efforts. Project 908, a fully owned government effort to manufacture semiconductors, began with an infusion of foreign technology (in this case, the purchase of a semiconductor fab from Lucent, whose executives say they were happy to be rid of it), but the plant was not competitive, not using the fab to its full capacity, and produced obsolescent microprocessors. Only after the company was "privatized" (i.e., government interventions in operations markedly decreased) did it become competitive. Removing "direct and operational interference" by government entities was a necessary first step for development.[52] The problems of two prominent IT projects suggest that government management and direction is not the source of China's IT success.

A sense of the range of Chinese software products can be gained from a leaked list from the Ministry of Finance (which is responsible for government procurement) on software approved for purchase by government ministries. Of the 73 listed software products approved for purchase, 10 were operating systems. The first 8 were Chinese Linux products (starting with Red Flag), and the last 2 were Windows variants (3 of the only 4 Western products approved for purchase were from Microsoft). In security software, only Chinese products are listed as approved. These include 5 antivirus products, 5 firewalls, 13 IDS and network management products, and 3 authentication products.[53]

China also plans to make its own information security software (firewalls, authentication programs, and intrusion detection systems) as part of the larger effort to develop a competitive software industry. Information security technologies are one of four priority programs for the current 863 project plans. In 2001, the 863 Program funded 56 information security projects for secure infrastructure, monitoring and control, and application security. In 2001 CERN, the China Education and Research Network, identified information security as first priority for new high-tech projects carried out under the 863 Program.[54] The

---

[52] Vipin Gupta and Jifu Wang, "From Corporate Crisis to Turnaround in East Asia: A Study of China Huajing Electronics Group Corporation," *Asia Pacific Journal of Management* 21: 213–233.

[53] United States Information Technology Office, "Recommended Software Products for Government Procurements (Partial); Unofficial USITO Translation," http://www.usito.org/uploads/177/254/SoftwareProcurementList.doc.

[54] See http://www.863.org.cn/english/annual_report/annual_repor_2001/200210090014.html; Du Mingha, "The 863 Program Blueprinting China's Future, China Education and Research Network, 2000, http://www.edu.cn/20010101/22306.shtml.

863 Program's 2001 Annual Report (the last published) identified information security as an "emergency subject" with these objectives:

- Create research programs for information security technologies to provide technologies and platforms for a security assurance architecture;

- Develop information security products and systems;

- Promote Chinese information security standards;

- Build "strong innovative abilities" in the field of information security.[55]

A Ministry of Public Safety survey found 200 information security companies offering 335 products at the end of 2000. The survey found another 350 companies doing research on security software. Western security software companies that have examined products say that these companies are generally not globally competitive. This may have helped lead to a government decision in 1998 to open the Chinese information security market to foreign-made software (subject to review and approval of the product by the Ministry of Public Safety).[56]

Lists of approved software for government acquisition aim both to help China's software industry and to restrict the use of foreign software that the Chinese fear may contain vulnerabilities inserted by foreign governments.[57] Business representatives say they are not surprised when they learn that the State Council or ministries issue such lists or acquisitions directives. Ministries (for IT, the Ministry of Commerce and the Ministry of Information Industries) also issue competing rulings to give companies associated with them an advantage not only over foreign firms but also over firms with ties to other ministries. Western companies report that the complex and overlapping system of regulation and regulatory agencies complicates the introduction of new IT products developed outside of China, particularly as administration and adjudication of regulations and directives vary from ministry to ministry. Chinese firms also complain about the multiple reviews and approvals required for products.

China's efforts to regulate encryption software demonstrate these tendencies. In the late 1990s, when Western law enforcement representatives raised the problems posed by the widespread use of commercial encryption, they found a receptive audience in China, and the Chinese may have been inspired in part to create the State Encryption Management Commission (SEMC) by the plans of the United States and other nations to control encryption. Western law enforcement alerted and sensitized the Chinese to the problems that the widespread use of encryption on public networks could cause, but China's perception of the potential problems went far beyond the concerns of their Western interlocutors.

---

[55] See http://www.863.org.cn/english/annual_report/annual_repor_2001/200210090014.html.

[56] Announced by Tian Qiyu, vice minister of public security, at the China International Information Security Exhibition in November 2000.

[57] ZDNet, China Blocks Foreign Software Use in Gov't," September 3, 2003, http://www.zdnet.com.au/news/software/0,2000061733,20277354,00.htm.

This concern led to an ill-conceived effort by Beijing to license foreign-produced encryption software. In October 1999, the State Council issued Directive Number 273, "Regulations for Managing Commercial Encryption," which authorized the new State Encryption Management Commission to require licensing of encryption products. U.S. officials and company representatives involved in the discussions of the new regulations say that China's motives were to prevent the development of communications channels that could not be monitored and to avoid the introduction of potentially compromised Western products. Because Chinese officials underestimated the amount of resources required to register encryption products, SEMC, a relatively small office, was overwhelmed by applications. Foreign companies and governments also made concerted protests against the new regulations, which, as a result, were suspended after a few months (March 2000).[58]

Although the regulations on encryption were withdrawn, the thinking behind them continues to shape policy. The Chinese government continues to certify foreign-origin software for domestic use, and security products (such as firewalls or intrusion-detection systems) receive additional scrutiny. It is not clear how strenuously certification requirements are enforced. Anecdotal evidence from Chinese and Western sources suggests that enforcement is periodic or sporadic. Western companies routinely comply to avoid the risk of future problems, and this gives the Chinese government access to considerable information on security software. The effect of this certification process is to reassure the government on its national security concerns and, perhaps, provide insight into foreign intellectual property.[59]

The electronic games market is the most dynamic sector in China's software industry. Many Chinese are passionate Internet gamers, something that prompts a degree of concern from the governments and explains some of the repeated efforts to regulate Internet cafes. Although Korean games at first dominated the China market (Korea has a large population of enthusiastic gamers), Chinese game developers have come into their own. Because of piracy concerns, most of their products are offered as online games (e.g., a participant does not buy a disk with the game, but instead pays a subscription fee to access the game over the Internet). Games must be first approved by the Ministry of culture to ensure that their content is not "violent, sexually explicit or harmful to national security."[60]

China's IP protection problems are an obstacle to the development of a strong domestic software industry. Much of the commercial software used in China is pirated. Although much of the attention to the piracy problem has come from foreign firms worried about revenue loss, the current situation deters investors and software developers from entering the market. Government incentives can only

---

[58] Interview with officials involved in the negotiations.

[59] See, for example, the U.S. firm Symark Corporation's press release on certification from March 2003 at http://www.symark.com/pr/2003/2003_03_27_symark_certified_in_china.htm.

[60] *People's Daily Online,* "China Says Games Must Get Approval," June 2, 2004, http://english.people.com.cn/200406/02/eng20040602_145171.html.

compensate in part for the limiting effect of piracy, and the growth of a Chinese software industry will depend on improvements in IP law and enforcement.

Computers

"Hardware must combine with software to power further development."

Sun Pishu, chairman, Langchao Qilu Software

China's computer companies are the test for the viability of the government efforts to create semiconductor and software industries. Given the Chinese penchant for prestige projects, it is not surprising that government entities focused for years on developing a Chinese supercomputer. China's efforts to build supercomputers have been aided by the development of "cluster" technology, where hundreds or thousands of commercial microprocessors are coupled together to provide high-performance computing. The spread of cluster supercomputers has been accelerated by the Internet availability of the software needed to run a cluster. China's most powerful computers are built using processors from American companies, and China is just one of many countries that take advantage of clustering technologies to create high-performance computing capabilities. However, supercomputer production is dominated by U.S. and Japanese firms. Chinese producers (Lenovo and Dawning) have built less than 1 percent of the 500 fastest supercomputers in the world.[61] The real action in computers lies elsewhere—in the PC and server industry.

Unlike supercomputer efforts, which are carried out by government institutions, PCs are made by private (or privatized) companies. PCs and servers made in China are commercial products; Chinese-made supercomputers are not. The PC industry demonstrates the central anomaly for China's IT industry. Most of the world's PCs and laptops are made in China, but they are often made by foreign firms using foreign components, software, and, most important, foreign intellectual property. China sees its chief challenge as transitioning from an industry that exists to build IT for multinational firms to an industry based on indigenously owned production using indigenous IP. This will not be easy, as the current system for producing PCs and laptops—assembling foreign components in China for a Taiwanese company under contract to a U.S. or Japanese company—is ruthlessly efficient.

Chinese computer makers face a market already occupied by large multinationals with staggeringly efficient international supply chains and sales networks. These companies already compete and collaborate with suppliers in many different countries and with each other. Taiwanese companies made about 80 percent of the global supply of laptops, using semiconductors and software from the United States and other Pacific Rim countries (though these components

---

[61] *Top 500 Supercomputer Sites,* Statistics type: Vendors, November 2005, http://www.top500.org/stats.

are themselves manufactured at facilities in other countries, ranging from Ireland to Singapore). For example, one Taiwanese company, Quanta, assembled about a quarter of all laptops sold around the world in 2004. Quanta moved much of its assembly work to China in the 1990s to take advantage of subsidies and lower costs, but expanded its R&D and design efforts in Taiwan. Ninety percent of Quanta's output comes from its factories in Shanghai. Encouraging relocation by companies like Quanta as the foundation for indigenous production, however, will not be easy.[62]

In addition, Chinese PC makers face competition from the "white-box" manufacturers—small computer makers who assemble cheap PCs and who account for more than a quarter of all PC sales. Some small Chinese white-box firms buy scrap computers in bulk from Western countries and reassemble or refurbish them for resale (often with pirated software) at very low prices.

Lenovo Computer Group (the company formerly named Legend) is China's premier computer maker. Lenovo's history helps to illustrate common elements in the strategies of China's leading IT companies for achieving a global presence. They rely extensively on partnerships with foreign firms for expertise and technology. They have strong support from government entities that can give them an edge in the domestic market, but have largely won independence from government control. Finally, they aim at winning a dominant position in sales within China and then make this the foundation for global expansion. These strategies could be found among computer makers in any developing nation, but Chinese companies may have an advantage given the extent to which multinationals use China as a manufacturing base to service the global market.

Lenovo, like other leading Chinese computer companies such as Langchao and TCL, started as a state-owned entity that commercialized itself and is now the most ambitious of these spin-offs. Researchers from the Chinese Academy of Sciences founded Legend in 1984, and CAS remains Legend's majority shareholder. At first Lenovo (as Lenovo) focused on selling foreign-made products. In 1990, it decided to make its own brand of PC, and after initial problems (the first Legend-designed computers were not very good), Lenovo worked with its foreign partners to come up with a low-cost, reliable PC. Several large multinational firms, such as Intel, Hitachi, and HP, helped Lenovo to integrate CPUs and software. Lenovo sold its PCs in China, gaining market share rapidly in its first years because of low-cost government connections, close attention to its national sales and support network, and a marketing campaign well attuned to Chinese tastes (auspiciously named computers in appealing colors). Some Chinese consumers preferred Lenovo because its network provided support in Chinese. Lenovo says it has 5,000 retail outlets in China.[63]

---

[62] See http://investintaiwan.nat.gov.tw/en/news/200502/2005021001.html; Andrew Tanzer, "Made in Taiwan," *Forbes*, April 2, 2001, http://www.forbes.com/global/2001/0402/024.html; Jason Dean and Pui-Wing Tam, "The Laptop Trail," *Wall Street Journal*, June 9, 2005, p. B1.
[63] Author's interviews with U.S. and Chinese executives; see http://www.cfoasia.com/archives/9905-22.htm; "Lenovo Announces New Wide Screen Thinkpads," *CNet News*,

After reaching the top of the China PC market, Legend decided to diversify both its product line and its markets. It adopted a new name, Lenovo, for this global effort (the Legend brand was already taken by another company). Two years later, it was forced to scale back its expansion. Lenovo's chief weaknesses were its lack of international marketing and management skills needed to sustain a global presence. In the interim, Lenovo faced renewed competition from leading U.S. PC makers who sought to expand their market share in China. Lenovo temporarily returned to a business strategy focused on China and saw its profits and market share recover by 2005.[64]

Lenovo renewed its efforts to become a global company in 2005 by buying IBM's PC and laptop division. Executives involved in the sale say that Lenovo's goal was not access to computer technology but access to IBM's global marketing skills and international brand. Lenovo hopes that by purchasing IBM's computer division, it will gain the business skills needed for a global presence. In the first year after its IBM acquisition, however, Lenovo's market share remained flat, and the new company stayed in third place, selling about half as many computers worldwide as either Dell or Hewlett-Packard. In response, Lenovo hired a new CEO—an American who was the president of Dell's China operations.[65]

Lenovo is one of the few companies that are fulfilling China's national goal of becoming global competitors in the IT market. These include SMIC for semiconductors and HuaWei for telecom. Langchao and Great Wall may join Lenovo in computers. These companies face two challenges: winning a greater share of the growing domestic market from foreign producers and establishing themselves as global competitors. The likely outcome of China's determination to have a national champion (or champions) in computer manufacturing is that a few companies, such as Lenovo, will achieve global presence and join the ranks of leading global IT firms. The others will be absorbed through industry consolidation if they are not carried by Chinese government subsidies. The goal, however, of a first-class computer made in China entirely by a Chinese company using only indigenous parts and software remains distant.

September 15, 2005, http://news.com.com/Lenovo+announces+new+wide-screen+ThinkPads/2100-1005_3-5862939.html?tag=nl.

[64] "Lenovo Chairman Liu Chuanzhi: 'We Have Decided to Refocus on the PC Business,'" *Knowledge@Wharton,* http://knowledge.wharton.upenn.edu/index.cfm?fa=viewArticle&id=1035.

[65] Gary McWilliams, "PC Shipments Jump," *Wall Street Journal,* January 19, 2006, http://online.wsj.com/article/SB113762996796250289.html?mod=home_whats_news_asia.

Investment and Overcapacity

"If China started to build a huge number of new fabs, we would have an industry where a majority of players would lose a lot of money."

Robert Tsao, chairman, United Microelectronics Corporation[66]

There is now a powerful IT industry in China, created by the intersection of government spending and global market forces, but the pace of growth raises the question as to whether the rapid expansion of the industry is sustainable. Growth has been rapid in good measure because of government funding for new companies. Two key questions are what happens to the IT sector in China if this spending shrinks and whether the government has overinvested in the IT sector.

Overinvestment can occur when governments or individuals decide to invest in a sector because of an inaccurate estimate of future demand. It can also occur when governments identify an economic or industrial goal and allocate funds to it beyond what is needed or what the market would support. The result can be a situation where the supply of a particular good or service is far greater than the demand.

Recent examples of overinvestment include the immense spending by telecom companies on unwanted capacity or the investment in new firms supplying Internet services. Historical examples include railroads, where companies laid far more track than was required. Government-directed programs can be especially prone to overinvestment, as political directions supplant market signals on where best to spend money. Both Korea and Japan saw government policy and political interference drive a misallocation of capital through strategies that, in the short term, produced rapid industrialization but in the last few years have worked to slow economic growth. In Korea and Japan, easy access to credit allowed less efficient firms to survive, draining resources from more productive activities and eventually leading to painful contractions in economic activity. The government investment in China's IT sector shares some similarities with these precedents.

For China, the risk of overinvestment springs from a number of sources. National policy has made the creation of the Chinese IT industry a priority, with the result that ministries and provincial or local governments invest more in this sector than they otherwise would have done. Government policy has also made credit easier to obtain for such investments, especially for enterprises where the state has an ownership role. Easy credit is a key ingredient for overinvestment; a boom mentality in cities like Shanghai is fed by the undeniable success of many Chinese companies but also by uncomplicated access to financing. At the same time, the percentage of nonperforming loans in the portfolios of Chinese banks

---

[66] See http://archives.cnn.com/2001/BUSINESS/asia/09/21/tw.umcthreats/.

may be as high as 45 percent, making most of them technically insolvent by Western standards.[67]

Centrally directed economies are more likely to mis-invest, and many spending decisions in China are still shaped by government policy. Weaknesses in the financial system and political interference in investment decisions constrain the ability of China's domestic capital market to accurately assess risk, and they increase the chance of overinvestment. Ministries and provincial and local governments provide funds or direct state-owned or controlled financial institutions to make loans. A strong element of central planning still exists, and Chinese government investments have not performed well in the past. These factors suggest that the Chinese IT sector may have unappreciated vulnerabilities that could affect its performance.

A tightening of credit in China could lead to a shakeout among IT companies. New Chinese IT companies could also go out of business if their access to government-provided credit ends (tighter credit triggered the collapse of the U.S. dot.com boom and accelerated problems in Japan and in the Korean Chaebol). Much of the current investment in IT in China is being made in anticipation of demand growth in the future. Many new Chinese IT companies will shrink or disappear if demand falls or if they are unable to supplant foreign competitors in supplying China's domestic IT market. As long as China's economy continues to grow, adding millions of new consumers to the market every year, and as long as the government continues to pump money into the IT sector, the consequences of overinvestment in IT can be avoided.

New production capabilities in China could result in intense global competition if Chinese domestic demand slackens. In this situation, Chinese firms might find themselves at a disadvantage. In the face of falling domestic demand, they would need to increase their sales to foreign markets. Few of these companies are ready to compete as global companies. Many lack the skills in international management, logistics, distribution, and marketing needed to run a global company. These are not skills that the Chinese government investment in technical education can provide. Chinese firms could obtain these skills by hiring foreigners, but they do not yet draw as heavily upon the international labor pool of business talent used in the United States and Europe and even in some larger Taiwanese firms.[68] Anecdotal evidence suggests that Indian firms have an advantage in attracting global talent (if only from the pool of Indians educated abroad) that may give India a relative advantage over China.

It is possible that the investment/incentive approach China has used to build an IT industry has reached its limits. It is an expensive policy, and some investments by regional governments seem driven more by a desire for prestige and hope than

---

[67] Huang Yasheng, Massachusetts Institute of Technology, Sloan School of Management, remarks at CSIS, May 5, 2004.

[68] "Chinese Entrepreneurs: On Their Way Back," *The Economist,* November 6, 2003; Embassy of China, "More Overseas Chinese Students Returning Home to Find Opportunities," November 16, 2003, http://www.china-embassy.org/eng/gyzg/t42338.htm.

by sound economic analysis. In recent years, Beijing has sought to cool investment in several sectors, and indications are that China is looking for new means to drive IT growth and give its IT companies an edge over foreign competitors. These measures could include exclusionary domestic standards, mandatory acquisitions requirements for government agencies (using, for example, a national security rationale that, under WTO rules, limits the ability of foreign producers to appeal the restriction), or aggressive use of antidumping regulations to damage foreign competitors. These approaches do not require the same level of government spending but still provide an advantage to domestic producers.

Conclusion

China has followed the precedent of other Asian countries in using preferences, incentives, and subsidies (public spending on infrastructure, human capital, low-cost financing for companies and entrepreneurs, trade and tax inducements) to attract foreign investment and develop a strong national IT sector. China's leaders hoped for both economic and security benefits from this effort. The economic benefits are clear. IT generates income, employment, and a deepening of human capital. In security, the long-term benefit to China lies in accelerated modernization and economic growth and, potentially, in the development of sophisticated IT skills that could be used to modernize China's military.

However, one important goal has not been achieved. Although government concerns are being supplanted by economic and business considerations as the balance between private sector and government shifts in China, the goal of improving information security by relying on domestically produced hardware and software remains an important motive for China's IT effort. The price of IT success for China was integration into a global production network and the close and essential foreign involvement. This integration undermines a strategy that relies on domestic production. In the long term, the security element of China's IT strategy is likely to fail. The fundamental premise for security in this strategy—that strong domestic industries not only benefit the economy but are also more trustworthy than foreign suppliers—no longer make sense in an integrated global industry and will not work.

China's IT security strategy comes from an older, pre-globalization approach based on a territorial concept of security and industrial policy still found in many countries. In this approach, borders were clear demarcations for trust. Key services and industries were state-owned or owned by national firms and therefore trustworthy. Global economic integration has eroded the territorial basis for trust and security. Trade agreements and technology make it easier for foreign nationals to invest, own companies, form partnerships, or provide services in another country. Globally integrated companies are more competitive, particularly in IT, as they have better access to ideas and markets. Firms that are not globally integrated are less competitive, lag technologically, and must be propped up with subsidies.

The territorial approach assumes that technology produced in China by a Chinese firm is more trustworthy (because foreign access to technology and production is limited). This assumption no longer holds. Even if the technology is manufactured in China, the global and collaborative nature of the IT business means that there will be foreign involvement. Industrial activities are now routinely conducted through international partnerships. Companies routinely collaborate with foreign partners to take advantage of specialization, to spread the risk of development, or to ensure market entry for new products. The leading Chinese IT companies have foreign partners and rely on foreign technology. A focus on country of origin for software and hardware is increasingly of uncertain value for improving security. China will be at a disadvantage in industrial and information security if it remains firmly wedded to a territorial approach.

China is not unique in this. Developing a new model for security that is not based on a reliance on national industries is a challenge for all nations. Nor, as some analyses seem to assume, is China unique or somehow immune from the pressures and risks that globalization creates for other countries. China's IT sector must be seen in the context of the larger global industry into which it is integrated.

In this global context, the government-driven investment and incentive model that has succeeded so well may no longer be effective. These policies appear to work best (at least initially) in developing economies where resources are underutilized and markets are inefficient. As the Chinese IT industry matures, it will face the same economic pressures in the IT sector that have led Western companies to come to China. China's firms will face the same competition for ideas, talent, and markets found elsewhere. Many of the new IT companies in China will not survive, but a few of these survivors will become global competitors on a par with their Japanese, American, and European counterparts.

China's efforts to build an IT industry began with the idea that its national security would improve if it could rely on domestic production. However, economic and commercial forces are submerging that security objective. For continued success in an integrated global economy, Beijing will need to cede even greater control over investment and operations to the private sector and to foreign companies. Even with a huge and talented labor pool, a giant domestic market, and a supportive government, China's IT industry will need to become more open and more internationally integrated if it is to remain competitive and continue to grow.

About the Author

James A. Lewis is a senior fellow and director of the CSIS Technology and Public Policy Program. Lewis, a former member of the U.S. Foreign Service and the Senior Executive Service, worked on foreign policy, national security, and technology-related issues at the Departments of State and Commerce. Since coming to CSIS, he has authored numerous publications, including *China as a Military Space Competitor* (2005), *Globalization and National Security* (2004), *Spectrum Management for the 21st Century* (2003), *Perils and Prospects for Internet Self-Regulation* (2002), *Assessing the Risk of Cyber Terrorism, Cyber War, and Other Cyber Threats* (2002), *Strengthening Law Enforcement Capabilities for Counterterrorism* (2001), and *Preserving America's Strength in Satellite Technology* (2001). His current research involves innovation, military space, and the global information technology industry. He received his Ph.D. from the University of Chicago in 1984.